FIBONACCI NUMBERS

by

N. N. Vorob'ev

➤ ➤ ➤

CONTENTS

FIBONACCI NUMBERS

A translation of "Chisla fibonachchi"
(Moscow—Leningrad, Gostekhteoretizdat, 1951)

Translated from the Russian by Halina Moss
Translation edited by Ian N. Sneddon
Additional editing by Robert R. Prechter, Jr.
Math typeset by Barbara Anderson
Text typeset by Paula Roberson
Cover and graphics by Roberta Machcinski

ISBN: 0-932750-03-6
Library of Congress Catalog Card Number: 83-60612

Editor's Note:
The notation for binomial coefficients is that used in the original text. For instance, on p. 18, the expression C_b^a is used, whereas common English usage would show ^bC_a.

FOREWORD

In elementary mathematics, there are many difficult and interesting problems not connected with the name of an individual, but rather possessing the character of "mathematical folklore". Such problems are scattered throughout the wide literature of popular (or, simply, entertaining!) mathematics, and often it is very difficult to establish the source of a particular problem.

These problems often circulate in several versions. Sometimes several such problems combine into a single more complex one; sometimes the opposite happens and one problem splits up into several simple ones. Thus it is often difficult to distinguish between the end of one problem and the beginning of another. We should consider that in each of these problems we are dealing with little mathematical theories, each with its own history, its own complex of problems and its own characteristic methods, all, however, closely connected with the history and methods of "great mathematics".

The theory of Fibonacci numbers is just such a theory. Derived from the famous "rabbit problem", going back nearly 750 years, Fibonacci numbers, even now, provide one of the most fascinating chapters of elementary mathematics. Problems connected with Fibonacci numbers occur in many popular books on mathematics, are discussed at meetings of school mathematical societies, and feature in mathematical competitions.

The present book contains a set of problems which were the themes of several meetings of the schoolchildren's mathematical club of Leningrad State University in the academic year 1949-50. In accordance with the wishes of those taking part, the questions discussed at these meetings were mostly number-theoretical, a theme which is developed in greater detail here.

This book is designed basically for pupils of
16 or 17 years of age in a high school. The
concept of a limit is encountered only in examples
7 and 8 in chapter III. The reader who is not
acquainted with this concept can omit these
without prejudice to his understanding of what
follows. That applies also to binomial
coefficients (I, example 8) and to trigonometry
(IV, examples 2 & 3). The elements which are
presented of the theories of divisibility and of
continued fractions do not presuppose any
knowledge beyond the limits of a school course.

Those readers who develop an interest in the
principle of constructing recurrent series are
recommended to read the small but comprehensive
booklet by A.I. Markushevich, "Recurrent
Sequences" (Vozvratnyye posledovatel' nosti)
(Gostekhizdat, 1950). Those who become interested
in facts relating to the theory of numbers are
referred to textbooks in this subject.

INTRODUCTION

<u>1</u>. The ancient world was rich in outstanding mathematicians. Many achievements of ancient mathematics are admired to this day for the acuteness of mind of their authors, and the names of Euclid, Archimedes and Hero are known to every educated person.

Things are different as far as the mathematics of the Middle Ages is concerned. Apart from Vieta, who lived as late as the sixteenth century, and mathematicians closer in time to us, a school course of mathematics does not mention a single name connected with the Middle Ages. This is, of course, no accident. In that epoch, the science developed extremely slowly and mathematicians of real stature were few.

Greater, then, is the interest of the work <u>Liber</u> <u>Abacci</u> ("a book about the abacus"), written by the remarkable Italian mathematician, Leonardo of Pisa, who is better known by his nickname Fibonacci (an abbreviation of filius Bonacci). This book, written in 1202, has survived in its second version, published in 1228.

<u>Liber</u> <u>Abacci</u> is a voluminous work, containing nearly all the arithmetical and algebraic knowledge of those times. It played a notable part in the development of mathematics in Western Europe in subsequent centuries. In particular, it was from this book that Europeans became acquainted with the Hindu (Arabic) numerals.

The theory contained in <u>Liber</u> <u>Abacci</u> is illustrated by a great many examples, which make up a significant part of the book.

Let us consider one of these examples, that which can be found on pages 123-124 of the manuscript of 1228:

> How many pairs of rabbits are born of one pair
> in a year?

This problem is stated in the form:

> Someone placed a pair of rabbits in a certain
> place, enclosed on all sides by a wall, to
> find out how many pairs of rabbits will be
> born there in the course of one year, it being
> assumed that every month a pair of rabbits
> produces another pair, and that rabbits begin
> to bear young two months after their own birth.

> As the first pair produces issue in the first
> month, in this month there will be 2 pairs.
> Of these, one pair, namely the first one,
> gives birth in the following month, so that in
> the second month there will be 3 pairs. Of
> these, 2 pairs will produce issue in the
> following month, so that in the third month 2
> more pairs of rabbits will be born, and the
> number of pairs of rabbits in that month will
> reach 5; of which 3 pairs will produce issue
> in the fourth month, so that the number of
> pairs of rabbits will then reach 8. Of these,
> 5 pairs will produce a further 5 pairs, which,
> added to the 8 pairs, will give 13 pairs in
> the fifth month. Of these, 5 pairs do not
> produce issue in that month but the other 8
> do, so that in the sixth month 21 pairs
> result. Adding the 13 pairs that will be born
> in the seventh month, 34 pairs are obtained;
> added to the 21 pairs born in the eighth month
> it becomes 55 pairs in that month; this, added
> to the 34 pairs born in the ninth month,
> becomes 89 pairs; and increased again by 55
> pairs which are born in the tenth month, makes
> 144 pairs in that month. Adding the 89
> further pairs which are born in the eleventh
> month, we get 233 pairs, to which we add,
> lastly, the 144 pairs born in the final
> month. We thus obtain 377 pairs. This is the
> number of pairs procreated from the first pair
> by the end of one year.

```
A pair
1

First   (Month)
2

Second
3

Third
5

Fourth
8

Fifth
13

Sixth
21

Seventh
34

Eighth -
55

Ninth
89

Tenth
144

Eleventh
233

Twelfth
377
```

Figure 1

From Figure 1 we see how we arrive at it: we
add to the first number the second one i.e., 1
and 2;   the second one to the third;   the
third to the fourth;   the fourth to the
fifth;   and in this way, one after another,
until we add together the tenth and the
eleventh numbers (144 and 233) and obtain the
total number of rabbits (377);   and it is
possible to do this in this order for an
infinite number of months.

<u>2</u>.   We now pass from rabbits to numbers and
examine the following numerical sequence

$$u_1, u_2, \ldots u_n,$$ \hfill (1)

in which each term equals the sum of two preceding
terms, i.e., for any $n \geq 2$,

$$u_n = u_{n-1} + u_{n-2}.$$ \hfill (2)

Such sequences, in which each term is defined as
some function of the previous ones, are
encountered often in mathematics, and are called
<u>recurrent</u> <u>sequences</u>. The process of successive
definition of the elements of such sequences is
itself called the <u>recurrence</u> <u>process</u>, and equation
(2) is called a <u>recurrence</u> <u>relation</u>. The reader
can find the elements of the general theory of
recurrent sequences in the book by Markushevich
mentioned in the Foreword.

We note that we cannot calculate the terms of
sequence (1) by condition (2) above.

It is possible to make up any number of
different numerical sequences satisfying this
condition. For example,

2, 5, 7, 12, 19, 31, 50, ...,
1, 3, 4, 7, 11, 18, 29, ...,
-1, -5, -6, -11, -17, ... and so on.

This means that for the unique construction of

sequence (1), the condition (2) is obviously
inadequate, and we must establish certain
supplementary conditions. For example, we can fix
the first few terms of sequence (1). How many of
the first terms of sequence (1) must we fix so
that it is possible to calculate all its following
terms, using only condition (2)?

We begin by pointing out that not every term
of sequence (1) can be obtained by (2), if only
because not all terms of (1) have two preceding
ones. For instance, the first term of the
sequence has no terms preceding it, and the second
term is preceded by only one. This means that in
addition to condition (2) we must know the first
two terms of the sequence in order to define it.

This is obviously sufficient to enable us to
calculate any term of sequence (1). Indeed, u_3
can be calculated as the sum of the prescribed
u_1 and u_2, u_4 as the sum of u_2 and the
previously calculated u_3, u_5 as the sum of the
previously calculated u_3 and u_4, and so on "in
this order up to an infinite number of terms".

Passing thus from two neighboring terms to the
one immediately following them, we can reach the
term with any required suffix and calculate it.

<u>3</u>. Let us now turn to the important particular
case of sequence (1), where $u_1 = 1$ and $u_2 = 1$. As
was pointed out above, condition (2) enables us to
calculate successively the terms of this series.
It is easy to verify that in this case the first
13 terms are the numbers 1, 1, 2, 3, 5, 8, 13, 21,
34, 55, 89, 144, 233, 377, which we already met in
the rabbit problem. To honor the author of the
problem, sequence (1) when $u_1 = u_2 = 1$ is called
the <u>Fibonacci</u> <u>sequence</u>, and its terms are know as
<u>Fibonacci</u> <u>numbers</u>.

Fibonacci numbers possess many interesting and
important properties, which are the subject of
this book.

I

THE SIMPLEST PROPERTIES OF FIBONACCI NUMBERS

$\underline{1}$. To begin with, we shall calculate the sum of the first n Fibonacci numbers. We shall show that

$$u_1 + u_2 + \ldots + u_n = u_{n+2} - 1. \tag{3}$$

Indeed, we have:

$$u_1 = u_3 - u_2,$$

$$u_2 = u_4 - u_3,$$

$$u_3 = u_5 - u_4$$

$$\ldots \ldots \ldots \ldots$$

$$u_{n-1} = u_{n+1} - u_n,$$

$$u_n = u_{n+2} - u_{n+1}.$$

Adding up these equations term by term, we obtain

$$u_1 + u_2 + \ldots + u_n = u_{n+2} - u_2,$$

and all that remains is to remember that $u_2 = 1$.

$\underline{2}$. The sum of Fibonacci numbers with odd suffixes

$$u_1 + u_3 + u_5 + \ldots + u_{2n-1} = u_{2n}. \tag{4}$$

To establish this equation, we shall write

$$u_1 = u_2,$$

$$u_3 = u_4 - u_2,$$

$$u_5 = u_6 - u_4,$$

$$\ldots \ldots \ldots \ldots$$

$$u_{2n-1} = u_{2n} - u_{2n-2}.$$

Adding these equations term by term, we obtain the required result.

<u>3</u>. The sum of Fibonacci numbers with even suffixes

$$u_2 + u_4 + \ldots + u_{2n} = u_{2n+1} - 1. \qquad (5)$$

From section 1 we have

$$u_1 + u_2 + u_3 + \ldots + u_{2n} = u_{2n+2} - 1;$$

subtracting equation (4) from the above equation, we obtain

$$u_2 + u_4 + \ldots + u_{2n} = u_{2n+2} - 1 - u_{2n} = u_{2n+1} - 1,$$

as was required.

Further, subtracting (5) from (4) term by term, we get

$$u_1 - u_2 + u_3 - u_4 + \ldots + u_{2n-1} - u_{2n} =$$

$$= -u_{2n-1} + 1. \qquad (6)$$

Now, let us add u_{2n+1} to both sides of (6):

$$u_1 - u_2 + u_3 - u_4 + \ldots - u_{2n} + u_{2n+1} =$$

$$= u_{2n} + 1. \qquad (7)$$

Combining (6) and (7), we get for the sum of Fibonacci numbers with alternating signs:

$$u_1 - u_2 + u_3 - u_4 + \ldots + (-1)^{n+1}u_n =$$

$$= (-1)^{n+1}u_{n-1} + 1. \qquad (8)$$

<u>4</u>. The formulas (3) and (4) were deduced by means of the term-by-term addition of a whole series of obvious equations. A further example of the application of this procedure is the proof of the formula for the sum of squares of the first n

Fibonacci numbers:

$$u_1^2 + u_2^2 + \ldots + u_n^2 = u_n u_{n+1}. \tag{9}$$

We note that

$$u_k u_{k+1} - u_{k-1} u_k = u_k(u_{k+1} - u_{k-1}) = u_k^2.$$

Adding up the equations

$$u_1^2 = u_1 u_2,$$

$$u_2^2 = u_2 u_3 - u_1 u_2,$$

$$u_3^2 = u_3 u_4 - u_2 u_3,$$

$$\ldots \ldots \ldots \ldots$$

$$u_n^2 = u_n u_{n+1} - u_{n-1} u_n$$

term by term, we obtain (9).

<u>5</u>.   Many relationships between Fibonacci numbers are conveniently proved with the aid of the method of induction.

The essence of the method of induction is as follows. In order to prove that a certain proposition is correct for any natural number, it is sufficient to establish:

(a) that it holds for the number 1;

(b) that from the truth of the proposition for an arbitrary natural number n follows its truth for the number n + 1.

Any inductive proof of a proposition true for any natural number consists, therefore, of two parts.

In the first part of the proof, the truth of the proposition is established for n = 1.   The truth of the proposition for n = 1 is sometimes called the <u>basis</u> <u>of</u> <u>induction</u>.

In the second part, the truth of the proposition is assumed for a certain arbitrary (but fixed) number n, and from this assumption, often called the inductive assumption, the deduction is made that the proposition is also true for the number n + 1. The second part of the proof is called the <u>inductive transition</u>.

The detailed presentation of the method of induction and numerous examples of the application of different forms of this method can be found in I.S. Sominskii, "The Method of Mathematical Induction". Thus, in particular, the version of the method of induction with the inductive transition "from n and n + 1 to n + 2" employed by us below is given in Sominskii's book on page 9 and is illustrated there on page 16 by problems 18 and 19.

We prove by induction the following important formula:

$$u_{n+m} = u_{n-1}u_m + u_n u_{m+1}. \tag{10}$$

We shall carry out the proof of this formula by induction on m. For m = 1, this formula takes the form

$$u_{n+1} = u_{n-1}u_1 + u_n u_2 = u_{n-1} + u_n,$$

which is obviously true. For m = 2, formula (10) is also true, because

$$u_{n+2} = u_{n-1}u_2 + u_n u_3 = u_{n-1} + 2u_n =$$

$$= u_{n-1} + u_n + u_n = u_{n+1} + u_n.$$

Thus the basis of the induction is proved. The inductive transition can be proved in this form: supposing formula (10) to be true for m = k and for m = k + 1, we shall prove that it also holds when m = k + 2.

Thus, let

$$u_{n+k} = u_{n-1}u_k + u_n u_{k+1}$$

and

$$u_{n+k+1} = u_{n-1}u_{k+1} + u_n u_{k+2}.$$

Adding the last two equations term by term, we obtain

$$u_{n+k+2} = u_{n-1}u_{k+2} + u_n u_{k+3}.$$

and this was the required result.

Putting $m = n$ in formula (10), we obtain

$$u_{2n} = u_{n-1}u_n + u_n u_{n+1},$$

or

$$u_{2n} = u_n(u_{n-1} + u_{n+1}). \tag{11}$$

From this last equation, it is obvious that u_{2n} is divisible by u_n. In the next chapter we shall prove a much more general result.

Since

$$u_n = u_{n+1} - u_{n-1},$$

formula (11) can be rewritten thus:

$$u_{2n} = (u_{n+1} - u_{n-1})(u_{n+1} + u_{n-1}),$$

or

$$u_{2n} = u_{n+1}^2 - u_{n-1}^2,$$

i.e., the difference of the squares of two Fibonacci numbers whose positions in the sequence differ by two is again a Fibonacci number.

Similarly (taking $m = 2n$), it can be shown that

$$u_{3n} = u_{n+1}^3 + u_n^3 - u_{n-1}^3.$$

6. The following formula will be found useful in what follows:

$$u_{n+1}^2 = u_n u_{n+2} + (-1)^n. \tag{12}$$

Let us prove it by induction over n. For n=1, (12) takes the form

$$u_2^2 = u_1 u_3 - 1,$$

which is obvious.

We now suppose formula (12) proved for a certain n. Adding $u_{n+1}u_{n+2}$ to both sides of it, we obtain

$$u_{n+1}^2 + u_{n+1}u_{n+2} = u_n u_{n+2} + u_{n+1}u_{n+2} + (-1)^n$$

or

$$u_{n+1}(u_{n+1} + u_{n+2}) = u_{n+2}(u_n + u_{n+1}) + (-1)^n$$

or

$$u_{n+1}u_{n+3} = u_{n+2}^2 + (-1)^n.$$

or

$$u_{n+2}^2 = u_{n+1}u_{n+3} + (-1)^{n+1}$$

Thus, the inductive transition is established and formula (12) is proved for any n.

7. In a similar way, it is possible to establish the following properties of Fibonacci numbers:

$$u_1 u_2 + u_2 u_3 + u_3 u_4 + \ldots + u_{2n-1}u_{2n} = u_{2n}^2,$$

$$u_1 u_2 + u_2 u_3 + u_3 u_4 + \ldots + u_{2n}u_{2n+1} = u_{2n+1}^2 - 1,$$

$$n u_1 + (n-1)u_2 + (n-2)u_3 + \ldots + 2u_{n-1} + u_n =$$

$$= u_{n+4} - (n+3).$$

The proofs are left to the reader.

<u>8</u>.   It   turns   out   that   there   is   a   connection
between the Fibonacci numbers and another set of
remarkable   numbers,   the   binomial   coefficients.
Let us set out the binomial coefficients in the
following triangle, called Pascal's triangle:

C_0^0

$C_1^0 \quad C_1^1$

$C_2^0 \quad C_2^1 \quad C_2^2$

$C_3^0 \quad C_3^1 \quad C_3^2 \quad C_3^3$

. . . .

i.e.,

```
          1
          1    1
          1    2    1
          1    3    3    1
          1    4    6    4    1
          1    5   10   10    5    1
          1    6   15   20   15    6    1
```
. . . .

The straight lines drawn through the numbers
of this triangle at an angle of 45 degrees to the
rows   we   shall   call   the   "rising   diagonals"   of
Pascal's   triangle.   For   instance,   the   straight
lines passing through numbers 1, 4, 3, or 1, 5, 6,
1, are rising diagonals.

We shall show that the sum of numbers lying
along a certain rising diagonal is a Fibonacci
number.

Indeed, the first and topmost rising diagonal
of   Pascal's   triangle   is   merely   1,   the   first
Fibonacci   number.   The   second   diagonal   also
consists of 1.

To prove the general proposition, it is sufficient to show that the sum of all numbers making up the (n–2)th and the (n–1)th diagonal of Pascal's triangle is equal to the sum of the numbers making up the nth diagonal.

On the (n–2)th diagonal we have the numbers

$$C_{n-3}^0, \ C_{n-4}^1, \ C_{n-5}^2, \ \ldots \ ,$$

and on the (n–1)th diagonal the numbers

$$C_{n-2}^0, \ C_{n-3}^1, \ C_{n-4}^2, \ \ldots$$

The sum of all these numbers can be written thus:

$$C_{n-2}^0 + (C_{n-3}^0 + C_{n-3}^1) + (C_{n-4}^1 + C_{n-4}^2) + \ldots \qquad (13)$$

But for binomial coefficients

$$C_{n-2}^0 = C_{n-1}^0 = 1$$

and

$$C_k^i + C_k^{i+1} = \frac{k(k-1) \ \ldots \ (k-i+1)}{1.2 \ \ldots \ .i} +$$

$$+ \frac{k(k-1) \ \ldots \ (k-i+1)(k-i)}{1.2. \ \ldots \ .i.(i+1)} =$$

$$= \frac{k(k-1) \ \ldots \ (k-i+1)}{1.2. \ \ldots \ .i} \left(1 + \frac{k-i}{i+1}\right) =$$

$$= \frac{k(k-1) \ \ldots \ (k-i+1)}{1.2. \ \ldots \ .i} \cdot \frac{i+1+k-i}{i+1} =$$

$$= \frac{(k+1)k(k-1) \ \ldots \ (k-i+1)}{1.2. \ \ldots \ .i.(i+1)} = C_{k+1}^{i+1}.$$

Expression (13) therefore equals

$$C_{n-1}^0 + C_{n-2}^1 + C_{n-3}^2 + \ldots \ ,$$

i.e., the sum of the numbers lying on the nth diagonal of the triangle.

From this proof and formula (3) we immediately get: The sum of all binomial coefficients lying above the nth rising diagonal of Pascal's triangle (inclusive of that diagonal) equals $u_{n+2} - 1$.

Making use of formulas (4), (5), (6) and similar ones, the reader can easily obtain further identities connecting Fibonacci numbers with binomial coefficients.

<u>9</u>. So far, we have defined Fibonacci numbers by a recurrence procedure, i.e., inductively, by their suffixes. It turns out, however, that any Fibonacci number can also be defined directly, as a function of its suffix.

To see this, we investigate various sequences satisfying the relationship (2). We shall call all such sequences solutions of equation (2).

In future we shall denote the sequences

$$v_1, v_2, v_3, \ldots,$$

$$v'_1, v'_2, v'_3, \ldots,$$

$$v''_1, v''_2, v''_3, \ldots,$$

by V, V' and V" respectively.

To begin with, we shall prove two simple lemmas.

<u>Lemma 1</u>. If V is the solution of equation (2) and c is an arbitrary number, then the sequence cV (i.e., the sequence cv_1, cv_2, cv_3, ...) is also a solution of equation (2).

<u>Proof</u>: Multiplying the relationship

$$v_n = v_{n-2} + v_{n-1}$$

term by term by c, we get

$$cv_n = cv_{n-2} + cv_{n-1},$$

as was required.

Lemma 2. If the sequences V' and V" are solutions of (2), then their sum V' + V" (i.e., the sequence $v'_1 + v''_1$, $v'_2 + v''_2$, $v'_3 + v''_3$, ...) is also a solution of (2).

Proof: From the conditions stated in the lemma, we have

$$v'_n = v'_{n-1} + v'_{n-2}$$

and

$$v''_n = v''_{n-1} + v''_{n-2}.$$

Adding these two equations term by term, we get

$$v'_n + v''_n = (v'_{n-1} + v''_{n-1}) + (v'_{n-2} + v''_{n-2}).$$

Thus, the lemma is proved.

Now, let V' and V" be two solutions of equation (2) which are not proportional. We shall show that any sequence V which is a solution of equation (2) can be written in the form

$$c_1V'_1 + c_2V''_2, \tag{14}$$

where c_1 and c_2 are constants. It is therefore usual to speak of (14) as the general solution of the equation (2).

First of all, we shall prove that if solutions of (2) V' and V" are not proportional, then

$$\frac{v'_1}{v''_1} \neq \frac{v'_2}{v''_2}. \tag{15}$$

The proof of (15) is carried out by assuming the opposite.

For solutions V' and V'' of (2) which are not proportional, let

$$\frac{v'_1}{v''_1} = \frac{v'_2}{v''_2} .$$

(16)

On writing down the derived proportion, we get

$$\frac{v'_1 + v'_2}{v''_1 + v''_2} = \frac{v'_2}{v''_2}$$

or, taking into account that V' and V'' are solutions of equation (2),

$$\frac{v'_3}{v''_3} = \frac{v'_2}{v''_2} .$$

Similarly, we convince ourselves (by induction) that

$$\frac{v'_3}{v''_3} = \frac{v'_4}{v''_4} = \ldots = \frac{v'_n}{v''_n} = \ldots$$

Thus, it follows from (16) that the sequences V' and V'' are proportional, which contradicts the assumption. This means that (15) is true.

Now, let us take a certain sequence V, which is a solution of the equation (2). This sequence, as was pointed out in section 2 of the Introduction, is fully defined if its two first terms, v_1 and v_2, are given.

Let us find such c_1 and c_2, that

$$c_1 v'_1 + c_2 v''_1 = v_1,$$

(17)

$$c_1 v'_2 + c_2 v''_2 = v_2.$$

Then, on the basis of lemmas 1 and 2, c_1V' + c_2V'' gives us the sequence V.

In view of condition (15), the simultaneous equations (17) are soluble with respect to c_1 and c_2, no matter what the numbers v_1 and v_2 are:

$$c_1 = \frac{v_1 v_2'' - v_2 v_1''}{v_1' v_2'' - v_1'' v_2'} , \qquad c_2 = \frac{v_1' v_2 - v_2' v_1}{v_1' v_2'' - v_1'' v_2'}$$

[By the condition (15), the denominator does not equal zero].

Substituting the values of c_1 and c_2 thus calculated in (14), we obtain the required representation of the sequence V.

This means that in order to describe all solutions of equation (2), it is sufficient to find any two solutions of it which are not proportional.

Let us look for these solutions among geometric progressions. In accordance with lemma 1, it is sufficient to limit ourselves to the consideration of only those progressions whose first term is equal to unity. Thus, let us take the progression

$$1, q, q^2, \ldots$$

In order that this progression should be a solution of (2), it is necessary that for any n, the equality

$$q^{n-2} + q^{n-1} = q^n$$

should be fulfilled. Or, dividing by q^{n-2},

$$1 + q = q^2.$$

The roots of this quadratic equation, i.e.,

$$\frac{1 + \sqrt{5}}{2} \quad \text{and} \quad \frac{1 - \sqrt{5}}{2} ,$$

will   be   the   required   common   ratios   of   the
progressions.   We   shall   denote   them   by   a   and   B
respectively.   Note that aB = −1.

We   have   thus   obtained   two   geometric
progressions   which   are   solutions   of   (2).
Therefore, all sequences of the form

$$c_1 + c_2, \; c_1 a + c_2 B, \; c_1 a^2 + c_2 B^2, \; \ldots \qquad (18)$$

are solutions of (2).   As the progressions found
by   us   have   different   common   ratios   and   are
therefore not proportional, formula (18) gives us
all solutions of equation (2).

In   particular,   for   certain   values   of c_1 and
c_2,   formula   (18)   should   also   give   us   the
Fibonacci series.   For this, as was pointed out
above, it is necessary to find c_1 and c_2 from
the equations

$$c_1 + c_2 = u_1$$

and

$$c_1 a + c_2 B = u_2,$$

i.e. from the simultaneous equations

$$c_1 + c_2 = 1,$$

$$c_1 \frac{1 + \sqrt{5}}{2} + c_2 \frac{1 - \sqrt{5}}{2} = 1.$$

Having solved them, we get

$$c_1 = \frac{1 + \sqrt{5}}{2\sqrt{5}}, \qquad c_2 = -\frac{1 - \sqrt{5}}{2\sqrt{5}},$$

whence

$$u_n = c_1 a^{n-1} + c_2 B^{n-1} =$$

$$= \frac{1 + \sqrt{5}}{2\sqrt{5}} \left(\frac{1 + \sqrt{5}}{2} \right)^{n-1} - \frac{1 - \sqrt{5}}{2\sqrt{5}} \left(\frac{1 - \sqrt{5}}{2} \right)^{n-1},$$

i.e.,

$$u_n = \frac{\left(\frac{1 + \sqrt{5}}{2}\right)^n - \left(\frac{1 - \sqrt{5}}{2}\right)^n}{\sqrt{5}} \qquad (19)$$

Formula (19) is called <u>Binet's formula</u> in honor of the mathematician who first proved it. Obviously, similar formulas can be derived for other solutions of (2). The reader should do it for the sequences introduced in section 2 of the Introduction.

<u>10</u>. With the help of Binet's formula, it is easy to find the sums of many series connected with Fibonacci numbers.

For instance, we can find the sum

$$u_3 + u_6 + u_9 + \ldots + u_{3n}.$$

We have

$$u_3 + u_6 + \ldots + u_{3n} =$$

$$= \frac{a^3 - B^3}{\sqrt{5}} + \frac{a^6 - B^6}{\sqrt{5}} + \ldots + \frac{a^{3n} - B^{3n}}{\sqrt{5}} =$$

$$= \frac{1}{\sqrt{5}}(a^3 + a^6 + \ldots + a^{3n} - B^3 - B^6 - \ldots - B^{3n}),$$

or, having summed the geometric progressions involved,

$$u_3 + u_6 + \ldots + u_{3n} = \frac{1}{\sqrt{5}}\left(\frac{a^{3n+3} - a^3}{a^3 - 1} - \frac{B^{3n+3} - B^3}{B^3 - 1}\right).$$

But

$$a^3 - 1 = a + a^2 - 1 = a + a + 1 - 1 = 2a,$$

and similarly, $B^3 - 1 = 2B$. Therefore,

$$u_3 + u_6 + \ldots + u_{3n} = \frac{1}{\sqrt{5}}\left(\frac{a^{3n+3} - a^3}{2a} - \frac{B^{3n+3} - B^3}{2B}\right),$$

or after cancellations,

$$u_3 + u_6 + \ldots + u_{3n} = \frac{1}{\sqrt{5}}\left(\frac{a^{3n+2} - a^2 - B^{3n+2} + B^2}{2}\right) =$$

$$= \frac{1}{2}\left(\frac{a^{3n+2} - B^{3n+2}}{\sqrt{5}} - \frac{a^2 - B^2}{\sqrt{5}}\right) =$$

$$= \frac{1}{2}\left(u_{3n+2} - u_2\right) = \frac{u_{3n+2} - 1}{2}.$$

11. As another example of the application of Binet's formula, we shall calculate the sum of the cubes of the first n Fibonacci numbers.

We note that

$$u_k^3 = \left(\frac{a^k - B^k}{\sqrt{5}}\right)^3 = \frac{1}{5}\left(\frac{a^{3k} - 3a^{2k}B^k + 3a^kB^{2k} - B^{3k}}{\sqrt{5}}\right) =$$

$$= \frac{1}{5}\left(\frac{a^{3k} - B^{3k}}{\sqrt{5}} - 3a^kB^k \frac{a^k - B^k}{\sqrt{5}}\right) = .$$

$$= \frac{1}{5}(u_{3k} - (-1)^k 3u_k) = \frac{1}{5}(u_{3k} + (-1)^{k+1} 3u_k).$$

Therefore,

$$u_1^3 + u_2^3 + \ldots + u_n^3 =$$

$$= \frac{1}{5}[(u_3 + u_6 + \ldots + u_{3n}) + 3(u_1 - u_2 + u_3 - \ldots + (-1)^{n+1} u_n)],$$

or, using formula (8) and the results of the preceding section,

$$u_1^3 + u_2^3 + \ldots + u_n^3 = \frac{1}{5}\left(\frac{u_{3n+2} - 1}{2} + 3[1 + (-1)^{n+1} u_{n-1}]\right) =$$

$$= \frac{u_{3n+2} + (-1)^{n+1} 6u_{n-1} + 5}{10}.$$

12. It is relevant to ask the question: how
quickly do Fibonacci numbers grow with increasing
suffix? Binet's formula gives us a sufficiently
full answer even to this question.

It is not hard to prove the following theorem.

Theorem. The Fibonacci number u_n is the nearest
whole number to the nth term a_n of the geometric
progression whose first term is $a/\sqrt{5}$ and whose
common ratio equals a.

Proof: Obviously it is sufficient to establish
that the absolute value of the difference between
u_n and a_n is always less than $1/2$. But

$$\left| u_n - a_n \right| = \frac{a^n - B^n}{\sqrt{5}} - \frac{a^n}{\sqrt{5}} = \left| \frac{a^n - a^n - B^n}{\sqrt{5}} \right| = \frac{|B|^n}{\sqrt{5}}.$$

As $B = -0.618...$, therefore $|B| < 1$, and that
means that for any n, $|B|^n < 1$ and even more so
(since $\sqrt{5} > 2$) $\frac{|B|}{5} < \frac{1}{2}$. The theorem is proved.

The reader who is acquainted with the theory
of limits will be able to show, by slightly
altering the proof of this theorem, that

$$\lim_{n \to \infty} \left| u_n - a_n \right| = 0.$$

Using this theorem, it is possible to
calculate Fibonacci numbers by means of
logarithmic tables.

For instance, let us calculate u_{14} (u_{14} is
the answer to the Fibonacci problem about the
rabbits):

$\sqrt{5} = 2.2361$, $\log \sqrt{5} = 0.34949$;

$a = \dfrac{1 + \sqrt{5}}{2} = 1.6180$, $\log a = 0.20898$;

$$\log \frac{a^{14}}{\sqrt{5}} = 14 \times 0.20898 - 0.34949 = 2.5762,$$

$$\frac{a^{14}}{\sqrt{5}} = 376.9.$$

The nearest whole number to 376.9 is 377; this is u_{14}.

When calculating Fibonacci numbers of very large suffixes, we can no longer calculate all the figures of the number by means of available tables of logarithms; we can only indicate the first few figures of it, so that the calculation turns out to be approximate.

As an exercise, the reader should prove that in the decimal system, u_n for $n \geq 17$ has no more than n/4 and no fewer than n/5 figures. And of how many figures will u_{1000} consist?

II

NUMBER-THEORETIC PROPERTIES OF
FIBONACCI NUMBERS

Before we continue the study of Fibonacci numbers, we shall remind the reader of some of the simplest facts from the theory of numbers.

$\underline{1}$. First, we shall indicate the process of finding the greatest common divisor of numbers a and b.

Suppose we divide a by b with a quotient equal to q_0 and a remainder r_1. Obviously, a $=$ $bq_0 + r_1$ and $0 \leq r_1 < b$. Note that if $a < b$, $q_0 = 0$.

Let us further divide b by r_1 and let us denote the quotient by q_1 and the remainder by r_2. Obviously $b = r_1 q_1 + r_2$, and $0 \leq r_2 < r_1$. Since $r_1 < b$, therefore $q_1 \neq 0$. Then, dividing r_1 by r_2, we shall find $q_2 \neq 0$ and r_3 such that $r_1 = q_2 r_2 + r_3$ and $0 \leq r_3 < r_2$. We proceed in this manner for as long as it is possible to continue the process.

Sooner or later our process must terminate, since all the positive whole numbers r_1, r_2, r_3, ... are different, and every one of them is smaller than b. That means that their number does not exceed b, and the process should terminate no later than at the bth step. But it can only terminate when a certain division proves to be carried out perfectly, i.e., the remainder turns out equal to zero and it will be impossible to divide anything by it.

The process thus described bears the name of Euclidean Algorithm. As a result of its application we obtain the following sequence of equations:

$$a = bq_0 + r_1,$$

$$b = r_1 q_1 + r_2,$$

$$r_1 = r_2 q_2 + r_3,$$

$$\cdots \cdots \cdots$$

$$r_{n-2} = r_{n-1} q_{n-1} + r_n,$$

$$r_{n-1} = r_n q_n.$$

$$(20)$$

Let us examine the last non-zero remainder r_n. Obviously r_{n-1} is divisible by r_n. Let us now take the last but one equation in (20). On its right-hand side both terms are divisible by r_n; therefore r_{n-2} is divisible by r_n. Similarly, we show step by step (induction) that r_{n-3}, r_{n-4}, ... and finally a and b are divisible by r_n. Thus, r_n is a common divisor of a and b. Let us show that r_n is the greatest common divisor of a and b. In order to do this, it is sufficient to show that any common divisor of a and b will also divide r_n.

Let d be a certain common divisor of a and b. From the first equation of (20), we notice that r_1 should be divisible by d. But in that case, on the basis of the second equation of (20), r_2 is divisible by d. Similarly (induction) we prove that d "goes into" r_3, ..., r_{n-1} and, finally, r_n.

We have thus proved that the Euclidean algorithm when applied to the natural numbers a and b really does lead to their greatest common divisor. This greatest common divisor of the numbers a and b is denoted by (a, b).

As an example, let us find $(u_{20}, u_{15}) = (6765, 610)$:

$$6765 = 610 \times 11 + 55,$$

$$610 = 55 \times 11 + 5,$$

$55 = 5 \times 11$.

Thus, $(u_{20}, u_{15}) = 5 = u_5$. The fact that
the greatest common divisor of two Fibonacci
numbers turned out to be again a Fibonacci number
is not accidental. It will be shown later that
that is always the case.

2. There is an analogy between Euclid's algorithm
and a process in geometry whereby the common
measure of two commensurable segments is found.

Indeed, let us examine two segments, one of
length a, the other of length b. Let us subtract
the second segment from the first as many times as
it is possible (if $b > a$, obviously we cannot do it
even once) and denote the length of the remainder
by r_1. Obviously $r_1 < b$. Now, let us subtract
from the segment of length b the segment of length
r_1 as many times as possible, and let us denote
the newly obtained remainder by r_2. Carrying on
in this manner, we obtain a sequence of remainders
whose lengths, evidently, decrease. Up to this
point, the resemblance to Euclid's algorithm is
complete.

Subsequently, however, an important difference
of the geometrical process from Euclid's algorithm
for natural numbers is revealed. The sequence of
remainders obtained from the subtraction of
segments might not terminate, as the process of
such subtraction can turn out to be capable of
being continued indefinitely. This will happen if
the chosen segments are incommensurable.

From the considerations in section 1, it
follows that two segments whose lengths can be
expressed by whole numbers are always
commensurable.

We now establish several simple properties of
the greatest common divisor of two numbers.

<u>3</u>.   (a, bc) is divisible by (a, b). Indeed, b,
and therefore bc, is divisible by (a, b); a is
divisible by (a, b) for obvious reasons. This
means, according to the proofs in section 1, that
(a, bc) is divisible by (a, b) also.

<u>4</u>.   (ac, bc) = (a, b)c

<u>Proof</u>:   Let the equations (20) describe the
process of finding (a, b). Multiplying each of
these equations by c throughout, we shall, as is
easily verified, obtain a set of equations
corresponding to the Euclidean algorithm as
applied to the numbers ac and bc. The last
non-zero remainder here will be equal to $r_n c$,
i.e., (a, b)c.

<u>5</u>.   If (a, c) = 1, then (a, bc) = (a, b). Indeed,
(a, bc) divides (ab, bc), according to section 3.
But (ab, bc) = (a, c)b = 1 x b = b, in view of
section 4. Thus, b is divisible by (a, bc). On
the other hand, (a, bc) divides a. By section 1,
this means that (a, bc) divides (a, b) also. And
since according to section 3, (a, b) divides (a,
bc) as well, then (a, b) = (a, bc).

<u>6</u>.   a is divisible by b only if (a, b) = b. This
is obvious.

<u>7</u>.   If c is divisible by b, then (a, b) = (a + c,
b).

<u>Proof</u>:   Suppose that the application of the
Euclidean algorithm to the numbers a and b leads
to the set of equations (20). Let us apply the
algorithm to the numbers a + c and b. Since c is
divisible by b, as given, we can put $c = c_1 b$.
The first step of the algorithm gives us the
equation

$$a + c = (q_0 + c_1)b + r_1.$$

The subsequent steps of this algorithm will give us consecutively the second, third, etc., equations of the set (20). The last non-zero remainder is still r_n, and this means that (a, b) = (a + c, b).

A useful exercise for the reader would be to prove this theorem on the sole basis of the results of sections 3-6, i.e., without a repeated reference to the idea of the Euclidean algorithm and to the set (20).

We now consider certain properties of Fibonacci numbers concerning their divisibility.

8. Theorem. If n is divisible by m, then u_n is also divisible by u_m.

Proof: Let n be divisible by m, i.e., let $n = mm_1$. We shall carry out the proof by induction over m_1. For $m_1 = 1$, n = m, so that in this case it is obvious that u_n is divisible by u_m. We now suppose that u_{mm_1} is divisible by u_m and consider $u_{m(m_1+1)}$. But $u_{m(m_1+1)} = u_{mm_1 + m}$ and, according to (10),

$$u_{m(m_1+1)} = u_{mm_1-1}u_m + u_{mm_1}u_{m+1}.$$

The first term of the right-hand side of this equation is obviously divisible by u_m. The second term contains u_{mm_1} as a factor, i.e., is divisible by u_m according to the inductive assumption. Hence, their sum, $u_{m(m_1+1)}$, is divisible by u_m as well. Thus the theorem is proved.

9. The topic of the arithmetical nature of Fibonacci numbers (i.e., the nature of their divisors) is of great interest.

We prove that for a compound n other than 4, u_n is a compound number.

Indeed, for such an n we can write $n = n_1 n_2$, where $1 < n_1 < n$, $1 < n_2 < n$ and either $n_1 > 2$ or $n_2 > 2$. To be definite, let $n_1 > 2$. Then, according to the theorem just proved, u_n is divisible by u_{n_1}, while $1 < u_{n_1} < u_n$, and this means that u_n is a compound number.

10.   Theorem.   Neighboring Fibonacci numbers are prime to each other.

Proof: Let u_n and u_{n+1} have a certain common divisor $d > 1$, in contradiction of what the theorem states. Then their difference $u_{n+1} - u_n$ should be divisible by d. And since $u_{n+1} - u_n = u_{n-1}$, then u_{n-1} should be divisible by d. Similarly, we prove (induction) that u_{n-2}, u_{n-3}, etc., and finally u_1, will be divisible by d. But $u_1 = 1$. Therefore it cannot be divided by $d > 1$. The incompatibility thus obtained proves the theorem.

11.   Theorem.   For any m, n $(u_m, u_n) = u_{(m,n)}$.

Proof: To be definite, we suppose $m > n$, and apply the Euclidean algorithm to the numbers m and n:

$m = n q_0 + r_1$, where $0 < r_1 < n$,

$n = r_1 q_1 + r_2$, where $0 < r_2 < r_1$,

$r_1 = r_2 q_2 + r_3$, where $0 < r_3 < r_2$,

. . . .

$r_{t-2} = r_{t-1} q_{t-1} + r_t$, where $0 < r_t < r_{t-1}$,

$r_{t-1} = r_t q_t$.

As we already know, r_t is the greatest common divisor of m and n.

Thus, $m = n q_0 + r_1$; this means that

$(u_m, u_n) = (u_{nq_0+r_1}, u_n),$

or, by equation (10),

$(u_m, u_n) = (u_{nq_0-1}u_{r_1} + u_{nq_0}u_{r_1+1}, u_n),$

or by sections 7 and 8,

$(u_m, u_n) = (u_{nq_0-1}u_{r_1}, u_n),$

or by sections 10 and 5,

$(u_m, u_n) = (u_{r_1}, u_n).$

Similarly, we prove that

$(u_{r_1}, u_n) = (u_{r_2}, u_{r_1}),$

$(u_{r_2}, u_{r_1}) = (u_{r_3}, u_{r_2}),$

. . . .

$(u_{r_{t-1}}, u_{r_{t-2}}) = (u_{r_t}, u_{r_{t-1}}).$

Combining all these equations, we get

$(u_m, u_n) = (u_{r_t}, u_{r_{t-1}}),$

and since r_{t-1} is divisible by r_t, $u_{r_{t-1}}$ is also divisible by u_{r_t}. Therefore $(u_{r_t}, u_{r_{t-1}}) = u_{r_t}$. Noting, finally, that $r_t = (m, n)$, we obtain the required result.

In particular, from the above proof we have a converse of the theorem in section 8: if u_n is divisible by u_m, then n is divisible by m. For if u_n is in fact divisible by u_m, then according to section 6,

$(u_n, u_m) = u_m.$ (21)

But we have proved that

$(u_n, u_m) = u_{(n,m)}.$ (22)

Combining (21) and (22) we get

$$u_m = u_{(n, m)},$$

i.e., m = (n,m), which means that n is divisible by m.

12. Combining the theorem in section 8 and the corollary to the theorem in section 11, we have: u_n is divisible by u_m if, and only if, n is divisible by m.

In view of this, the divisibility of Fibonacci numbers can be studied by examining the divisibility of their suffixes.

Let us find, for instance, some "signs of divisibility" of Fibonacci numbers. By "sign of divisibility", we mean a sign to show whether any particular Fibonacci number is divisible by a certain given number.

A Fibonacci number is even if, and only if, its suffix is divisible by 3.

A Fibonacci number is divisible by 3 if, and only if, its suffix is divisible by 4.

A Fibonacci number is divisible by 4 if, and only if, its suffix is divisible by 6.

A Fibonacci number is divisible by 5 if, and only if, its suffix is divisible by 5.

A Fibonacci number is divisible by 7 if, and only if, its suffix is divisible by 8.

The proofs of all these signs of divisibility and all similar ones can be carried out easily by the reader, with the help of the proposition put forward at the beginning of the section, and by considering the third, fourth, sixth, fifth, eighth, etc. Fibonacci numbers respectively.

At the same time, the reader should prove that no Fibonacci number exists that would give a remainder of 4 when divided by 8; also, that there are no odd Fibonacci numbers divisible by 17.

13. Let us now take a certain whole number m. If there exists even one Fibonacci number u_n divisible by m, it is possible to find as many such Fibonacci numbers as desired. For example, such will be the numbers u_{2n}, u_{3n}, u_{4n}, $\ldots$

It would therefore be interesting to discover whether it is possible to find at least one Fibonacci number divisible by a given number m. It turns out that this is possible.

Let U be the remainder of the division of u by m, and let us write down a sequence of pairs of such remainders:

$$[U_1, U_2], [U_2, U_3], [U_3, U_4], \ldots, [U_n, U_{n+1}], \ldots \tag{23}$$

If we regard pairs $[a_1, b_1]$ and $[a_2, b_2]$ as equal when $a_1 = a_2$ and $b_1 = b_2$, the number of different pairs of remainders of division by m equals m^2. If, therefore, we take the first $m^2 + 1$ terms of the sequence (23), there must be equal ones among them.

Let $[U_k, U_{k+1}]$ be the first pair that repeats itself in the sequence (23). We shall show that this pair is [1, 1]. Indeed, let us suppose the opposite, i.e., that the first repeated pair is the pair $[U_k, U_{k+1}]$, where k > 1. Let us find in (23) a pair $[U_j, U_{j+1}]$ (j > k) equal to the pair $[U_k, U_{k+1}]$. Since $u_{j-1} = u_{j+1} - u_j$ and $u_{k-1} = u_{k+1} - u_k$, and $U_{j+1} = U_{k+1}$ and $U_j = U_k$, the remainders of division of u_{j-1} and u_{k-1} by m are equal, i.e., $U_{j-1} = U_{k-1}$. However, it also follows that $[U_{k-1}, U_k] = [U_{j-1}, U_j]$, but the pair $[U_{k-1}, U_k]$ is situated in the sequence (23) earlier than $[U_k, U_{k+1}]$ and therefore $[U_k, U_{k+1}]$ is not the first pair that repeated itself, which contradicts

our premise. This means that the supposition k> 1
is wrong, and therefore k = 1.

    Thus [1, 1] is the first pair that repeats
itself in (23). Let the repeated pair be in the
tth place (in accordance with what was established
earlier we can regard $1 < t < m^2 + 1$), i.e.,
$[U_t, U_{t+1}] = [1, 1]$. This means that both
u_t and u_{t+1}, when divided by m, give 1 as a
remainder. It follows that their difference is
exactly divisible by m. But $u_{t+1} - u_t =$
u_{t-1}, so that the (t − 1)th Fibonacci number is
divisible by m.

We have thus proved the following theorem:

Theorem: Whatever the whole number m, at least
one number divisible by m can be found among the
first m^2 Fibonacci numbers.

    Note that this theorem does not state anything
about exactly which Fibonacci number will be
divisible by m. It only tells us that the first
Fibonacci number divisible by m should not be
particularly large.

III

FIBONACCI NUMBERS AND CONTINUED FRACTIONS

<u>1</u>. We consider the expression

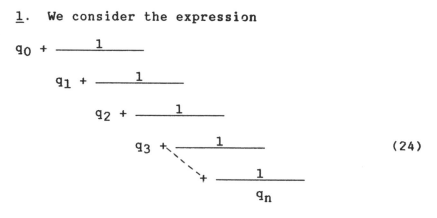

$$(24)$$

where q_1, q_2, ..., q_n are whole positive numbers and q_0 is a whole non-negative number. Thus, in contrast to the numbers q_1, q_2, ..., q_n, the number q_0 can equal zero. We shall keep this somewhat special position of the number q_0 in mind, and not mention it specially on each occasion.

The expression (24) is called a <u>continued fraction</u> and the numbers q_0, q_1, ..., q_n are called the <u>partial denominators</u> of this fraction.

Sometimes continued fractions are also known as chain fractions. They are of use in a wide assortment of mathematical problems. The reader who wants to study them in greater detail is referred to A.Ya. Khinchin, "Chain Fractions".

The process of transformation of a certain number into a continued fraction is called the <u>development</u> of this number into a continued fraction.

Let us see how we can find the partial denominators of such an expansion of the ordinary fraction a/b.

We consider the Euclidean algorithm as applied to the numbers a and b.

$$a = bq_0 + r_1,$$

$$b = r_1 q_1 + r_2,$$

$$r_1 = r_2 q_2 + r_3,$$

. . . .   (25)

$$r_{n-2} = r_{n-1} q_{n-1} + r_n,$$

$$r_{n-1} = r_n q_n.$$

The first of these equations gives us

$$\frac{a}{b} = q_0 + \frac{r_1}{b} = q_0 + \frac{1}{\frac{b}{r_1}}.$$

But it follows from the second equation of set (25) that

$$\frac{b}{r_1} = q_1 + \frac{r_2}{r_1} = q_1 + \frac{1}{\frac{r_1}{r_2}},$$

so that

$$\frac{a}{b} = q_0 + \frac{1}{q_1 + \frac{1}{\frac{r_1}{r_2}}}.$$

From the third equation of (25), we deduce

$$\frac{r_1}{r_2} = q_2 + \frac{r_3}{r_2} = q_2 + \frac{1}{\frac{r_2}{r_3}}$$

and therefore,

$$\frac{a}{b} = q_0 + \frac{1}{q_1 + \frac{1}{q_2 + \frac{1}{\frac{r_2}{r_3}}}}.$$

Continuing this process to the end (induction) we arrive, as is seen easily, at the equation

$$\frac{a}{b} = q_0 + \cfrac{1}{q_1 + \cfrac{1}{q_2 + \cfrac{\ddots}{\ + \cfrac{1}{q_n}}}} \qquad .$$

By the very sense of the Euclidean algorithm, $q_n > 1$. (If q_n were equal to unity, then r_{n-1} would equal r_n, and r_{n-2} would have been divisible by r_{n-1} exactly, i.e., the whole algorithm would have terminated one step earlier.) This means that in place of q_n, we can consider the expression $(q_n - 1) + 1/1$, i.e., consider (q_n-1) the last but one partial denominator, and 1 the last. Such a convention turns out to be convenient for what follows.

The Euclidean algorithm as applied to a given pair of natural numbers a and b is realized in a completely definite and unique way. The partial denominators of the development of a/b into a continuous fraction are also defined in a unique way by the system of equations describing this algorithm. Any rational fraction a/b, therefore, can be expanded into a continued fraction in one and only one way.

2. Let

$$w = q_0 + \cfrac{1}{q_1 + \cfrac{1}{q_2 + \cfrac{\ddots}{\ + \cfrac{1}{q_n}}}} \qquad (26)$$

be a certain continued fraction, and let us consider the following numbers:

$$q_0, \quad q_0 + \cfrac{1}{q_1}, \quad q_0 + \cfrac{1}{q_1 + \cfrac{1}{q_2}}, \quad \ldots$$

These numbers, written down in the form of ordinary simple fractions

$$\frac{P_0}{Q_0} = \frac{q_0}{1},$$

$$\frac{P_1}{Q_1} = q_0 + \frac{1}{q_1},$$

$$\frac{P_2}{Q_2} = q_0 + \cfrac{1}{q_1 + \cfrac{1}{q_2}},$$

. . . .

$$\frac{P_n}{Q_n} = w,$$

are called <u>convergent</u> fractions of the continued fraction w.

Note that the transition from $\frac{P_k}{Q_k}$ to $\frac{P_{k+1}}{Q_{k+1}}$ is realized by the replacement of the last of those partial denominators which took part in the construction of this convergent fraction, i.e., q_k, by $q_k + \dfrac{1}{q_{k+1}}$.

<u>3</u>.   The following lemma plays an important role in the theory of continued fractions.

<u>Lemma</u>.   For every continued fraction (26) the following relationships obtain:

$$P_{k+1} = P_k q_{k+1} + P_{k-1}, \tag{27}$$

$$Q_{k+1} = Q_k q_{k+1} + Q_{k-1}, \tag{28}$$

$$P_{k+1}Q_k - P_k Q_{k+1} = (-1)^k. \tag{29}$$

We prove all these equations simultaneously by induction over k.

We shall prove them first for k = 1.

$$\frac{P_1}{Q_1} = q_0 + \frac{1}{q_1} = \frac{q_0 q_1 + 1}{q_1}.$$

Since the numbers $q_0 q_1 + 1$ and q_1 are prime to each other, the fraction $\frac{q_0 q_1 + 1}{q_1}$ is reduced to its lowest terms.

The fraction P_1/Q_1 is in its lowest terms according to the definition, and equal fractions in their lowest terms have equal numerators and equal denominators. This means that $P_1 = q_0 q_1 + 1$ and $Q_1 = q_1$.

$$\frac{P_2}{Q_2} = q_0 + \frac{1}{q_1 + \frac{1}{q_2}} = \frac{q_0(q_1 q_2 + 1) + q_2}{q_1 q_2 + 1}. \qquad (30)$$

The greatest common divisor of the numbers $q_0(q_1 q_2 + 1) + q_2$ and $q_1 q_2 + 1$ equals $(q_2, q_1 q_2 + 1)$ on the basis of section 7 of II, and on the basis of the same proposition, it also equals $(q_2, 1)$, i.e., 1. This means that the fraction on the right-hand side in (30) is in its lowest terms, and therefore

$$P_2 = q_0(q_1 q_2 + 1) + q_2 = (q_0 q_1 + 1)q_2 + q_0 = P_1 q_2 + P_0$$

and

$$Q_2 = q_1 q_2 + 1 = Q_1 q_2 + Q_0.$$

The equation

$$P_2 Q_1 - P_1 Q_2 = (-1)^1$$

is easily verified.

The basis of the induction is thus proved.

Let us now suppose that the equations (27), (28) and (29) are true and let us consider the convergent fraction

$$\frac{P_{k+1}}{Q_{k+1}} = \frac{P_k q_{k+1} + P_{k-1}}{Q_k q_{k+1} + Q_{k-1}}.$$

The transition from $\dfrac{P_{k+1}}{Q_{k+1}}$ to $\dfrac{P_{k+2}}{Q_{k+2}}$ according to

the observation made above is brought about by the replacement of q_{k+1} in the expression for $\dfrac{P_{k+1}}{Q_{k+1}}$

by $q_{k+1} + \dfrac{1}{q_{k+2}}$;

since q_{k+1} does not come into the expressions for P_k, Q_k, P_{k-1}, Q_{k-1}, then

$$\frac{P_{k+2}}{Q_{k+2}} = \frac{P_k\left(q_{k+1} + \dfrac{1}{q_{k+2}}\right) + P_{k-1}}{Q_k\left(q_{k+1} + \dfrac{1}{q_{k+2}}\right) + Q_{k-1}},$$

or, remembering the inductive assumptions in (27) and (28),

$$\frac{P_{k+2}}{Q_{k+2}} = \frac{P_{k+1}q_{k+2} + P_k}{Q_{k+1}q_{k+2} + Q_k}. \qquad (31)$$

We now prove that the right-hand fraction in (31) is in its lowest terms. For this it is sufficient to prove that its numerator and denominator are mutually prime.

Let us suppose that the numbers $P_{k+1}q_{k+2} + P_k$ and $Q_{k+1}q_{k+2} + Q_k$ have a certain common divisor $d > 1$. The expression

$$(P_{k+1}q_{k+2} + P_k)Q_{k+1} - (Q_{k+1}q_{k+2} + Q_k)P_{k+1}$$

should then be divisible by d. But by the inductive assumption (29), this expression equals $(-1)^{k+1}$ and cannot be divided by d.

Thus the right-hand side of (31) is in its lowest terms, and (31) is therefore an equation between two fractions reduced to their lowest terms. This means that

$$P_{k+2} = P_{k+1}q_{k+2} + P_k$$

and that

$Q_{k+2} = Q_{k+1}q_{k+2} + Q_k$.

To complete the proof of the inductive transition, it remains to show that

$$P_{k+2}Q_{k+1} - P_{k+1}Q_{k+2} = (-1)^{k+1}. \tag{32}$$

But in view of what was proved above,

$$P_{k+2}Q_{k+1} - P_{k+1}Q_{k+2} =$$

$$= P_{k+1}q_{k+2}Q_{k+1} + P_kQ_{k+1} - P_{k+1}q_{k+2}Q_{k+1} - P_{k+1}Q_k,$$

and (32) follows directly from the inductive assumption (29). In this way, the inductive transition is established and the whole lemma is proved.

Corollary.

$$\frac{P_{k+1}}{Q_{k+1}} - \frac{P_k}{Q_k} = \frac{(-1)^k}{Q_kQ_{k+1}}. \tag{33}$$

Since partial denominators of continued fractions are positive whole numbers, it follows from the above lemma that:

$$P_0 < P_1 < P_2 < \cdots ,$$
$$Q_0 < Q_1 < Q_2 < \cdots . \tag{34}$$

This simple yet important observation will be made more exact later in the book.

4. We now apply the lemma of section 3 to describe all continued fractions with partial denominators equal to unity. For such fractions we have the following interesting theorem.

Theorem. If a continued fraction has n partial denominators and each of these partial denominators equals unity, the fraction equals $\frac{u_{n+1}}{u_n}$.

Proof: Let us denote the continued fraction with n unit partial denominators by a_n. Obviously, a_1, a_2, ..., a_n are consecutive convergent fractions of a_n.

Let

$$a_k = \frac{P_k}{Q_k}.$$

As

$$a_1 = 1 = \frac{1}{1}$$

and

$$a_2 = 1 + \frac{1}{1} = \frac{2}{1},$$

therefore, $P_1 = 1$, $P_2 = 2$. Further, $P_{n+1} = P_n q_{n+1} + P_{n-1} = P_n + P_{n-1}$. Therefore (compare I, section 8), $P_n = u_{n+1}$.

Similarly, $Q_1 = 1$, $Q_2 = 1$ and $Q_{n+1} = Q_n q_{n+1} + Q_{n-1} = Q_n + Q_{n-1}$, so that $Q_n = u_n$. This means that

$$a_n = \frac{u_{n+1}}{u_n}. \tag{35}$$

The reader should compare this result with formulas (12) and (29).

5. Suppose we are given two continued fractions w and w':

$$w = q_0 + \cfrac{1}{q_1 + \cfrac{1}{q_2 + \cdots}},$$

$$w' = q'_0 + \cfrac{1}{q'_1 + \cfrac{1}{q'_2 + \cdots}}$$

while

$$q_0' \geq q_0, \quad q_1' \geq q_1, \quad q_2' \geq q_2, \quad \cdots \tag{36}$$

Let us denote the convergent fractions of w by

$$\frac{P_0}{Q_0}, \quad \frac{P_1}{Q_1}, \quad \frac{P_2}{Q_2}, \quad \cdots$$

and the convergent fractions of w' by

$$\frac{P_0'}{Q_0'}, \quad \frac{P_1'}{Q_1'}, \quad \frac{P_2'}{Q_2'}, \quad \cdots$$

From the results of the lemma of section 3, it is easy to detect that in view of (36)

$$P_0' \geq P_0, \quad P_1' \geq P_1, \quad P_2' > P_2, \quad \cdots$$

and

$$Q_0' \geq Q_0, \quad Q_1' \geq Q_1, \quad Q_2' \geq Q_2, \quad \cdots$$

Obviously, the smallest value of any partial denominator is unity. This means that if all the partial denominators of a certain continued fraction are unity, the numerators and denominators of its convergent fractions increase more slowly than those of the convergent fractions of any other continued fraction.

Let us estimate to what extent this increase is slowed down. Obviously, discounting the continued fractions whose partial denominators are unity, the slowest to increase are the numerators and denominators of the convergent fractions of that continued fraction one of whose partial denominators is 2 and the remaining ones unity. Such continued fractions are also connected with Fibonacci numbers as shown by the following lemma:

Lemma. If the continued fraction w has as its partial denominators the numbers q_0, q_1, q_2, ..., q_n, while

$$q_0 = q_1 = q_2 = \cdots = q_{i-1} = q_{i+1} = \cdots =$$

$= q_n = 1$, $q_i = 2$ $(i \neq 0)$

then

$$w = \frac{u_{i+1}u_{n-i+3} + u_i u_{n-i+1}}{u_i u_{n-i+3} + u_{i-1}u_{n-i+1}}.$$

Proof of this lemma is carried out by induction over i. If i = 1, then for any n,

$$w = 1 + \cfrac{1}{2 + \cfrac{1}{1 + \begin{array}{c} \\ \ddots \end{array} + \cfrac{1}{1}}}$$
n-1 partial denominator

or, in view of what was proved at the beginning of this section,

$$w = 1 + \cfrac{1}{2 + \cfrac{1}{a_{n-1}}} = 1 + \cfrac{1}{2 + \cfrac{u_{n-1}}{u_n}} = 1 + \cfrac{1}{\cfrac{2u_n + u_{n-1}}{u_n}} =$$

$$= 1 + \cfrac{u_n}{u_{n+2}} = \frac{u_{n+2} + u_n}{u_{n+2}},$$

or, putting $u_0 = 0$,

$$w = \frac{u_2 u_{n+2} + u_1 u_n}{u_1 u_{n+2} + u_0 u_n}.$$

Thus, the basis of induction has been proved.

Let us now suppose that for any n,

$$\left\{ 1 + \cfrac{1}{1 + \begin{array}{c} \\ \ddots \end{array} + 1 + \cfrac{1}{2 + \cfrac{1}{a_{n-i}}}} \right\} =$$
i partial denominators

$$= \frac{u_{i+1}u_{n-i+3} + u_i u_{n-i+1}}{u_i u_{n-i+3} + u_{i-1}u_{n-i+1}}. \tag{37}$$

Let us take the continued fraction

$$\text{i+1 partial denominators} \begin{cases} 1 + \cfrac{1}{1 + } \\ + 1 + \cfrac{1}{2 + \cfrac{1}{a_{n-i-1}}} \end{cases}$$

It can obviously be considered thus:

$$\text{i partial denominators} \begin{cases} \overbrace{\cfrac{1 + 1}{1 + }}^{\cdots\cdots\cdots\cdots\cdots\cdots\cdots\cdots} \\ + 1 + \cfrac{1}{2 + \cfrac{1}{a_{n-i-1}}} \end{cases} \tag{38}$$

The continued fraction below the dotted line in (38) is, by (37), equal to

$$\frac{u_{i+1}u_{n-i+2} + u_i u_{n-i}}{u_i u_{n-i+2} + u_{i-1}u_{n-i}}.$$

The whole fraction (38) therefore equals

$$1 + \cfrac{1}{\cfrac{u_{i+1}u_{n-i+2} + u_i u_{n-i}}{u_i u_{n-i+2} + u_{i-1}u_{n-i}}} =$$

$$= \frac{(u_i + u_{i+1})u_{n-i+2} + (u_{i-1} + u_i)u_{n-i}}{u_{i+1}u_{n-i+2} + u_i u_{n-i}} =$$

$$= \frac{u_{i+2}u_{n-i+2} + u_{i+1}u_{n-i}}{u_{i+1}u_{n-i+2} + u_i u_{n-i}}.$$

Thus, the inductive transition has been proved and so has the whole lemma.

Corollary. If not all the partial denominators of the continued fraction w are unity, $q_0 \neq 0$, and there are no less than n of these partial denominators, then, on writing w in the form of an ordinary fraction P/Q, we have

$$P \geq u_{i+1}u_{n-i+3} + u_i u_{n-i+1} > u_{i+1}u_{n-i+2} +$$

$$+ u_i u_{n-i+1} = u_{n+2},$$

and similarly,

$Q > u_{n+1}$.

A substantial role is played here, of course, by the lemma of section 3, on the basis of which we obtain only fractions in their lowest terms in the process of "contracting" a continued fraction into a vulgar one. Therefore no diminution of numerators and denominators of the fractions obtained due to "canceling" will take place.

6.   Theorem. For a certain a, the number of steps in the Euclidean algorithm applied to the numbers a and b equals n - 1 if b = u_n, and for any a it is less than n - 1 if b < u_n.

Proof: The first part of the theorem can be proved quite simply. It is sufficient to take as a the Fibonacci number following b, i.e., u_{n+1}. Then,

$$\frac{u_{n+1}}{u_n} = a_n.$$

The continued fraction a_n has n partial denominators, i.e., the number of steps of the Euclidean algorithm as applied to the numbers a and b equals n - 1.

To prove the second part of the theorem, we suppose the contrary, i.e., that the number of steps of this particular algorithm is not less than n - 1. Let us expand the ratio a/b into the continued fraction w. Obviously w will have no less than n partial denominators (in fact one more than the length of the Euclidean algorithm). As b is not a Fibonacci number, not all the partial denominators of w will be unity, and therefore, according to the corollary of the lemma in section 5, b > u_n, which contradicts the conditions of the theorem.

This theorem means that the Euclidean algorithm as applied to neighboring Fibonacci numbers is, in a sense, the "longest".

$\underline{7}$. We shall call the expression

$$q_0 + \cfrac{1}{q_1 + \cfrac{1}{q_2 + \cdots + \cfrac{1}{q_n + \cdots}}} \tag{39}$$

an infinite continued fraction.

The definitions and results of the preceding sections can be extended quite naturally to infinite continued fractions.

Let

$$\frac{P_0}{Q_0}, \frac{P_1}{Q_1}, \cdots, \frac{P_n}{Q_n}, \cdots \tag{40}$$

be a sequence (obviously an infinite one) of the convergent fractions of the fraction (39).

We shall show that this sequence has a limit.

With this aim in mind, we examine separately the sequences

$$\frac{P_0}{Q_0}, \frac{P_2}{Q_2}, \cdots, \frac{P_{2n}}{Q_{2n}}, \cdots \tag{41}$$

and

$$\frac{P_1}{Q_1}, \frac{P_3}{Q_3}, \cdots, \frac{P_{2n+1}}{Q_{2n+1}}, \cdots \tag{42}$$

From (33) and (34),

$$\frac{P_{2n+2}}{Q_{2n+2}} - \frac{P_{2n}}{Q_{2n}} = \frac{P_{2n+2}}{Q_{2n+2}} - \frac{P_{2n+1}}{Q_{2n+1}} + \frac{P_{2n+1}}{Q_{2n+1}} - \frac{P_{2n}}{Q_{2n}} =$$

$$= \frac{-1}{Q_{2n+2}Q_{2n+1}} + \frac{1}{Q_{2n+1}Q_{2n}} > 0$$

This means that the sequence (41) is an increasing one. In the same way, it follows from

$$\frac{P_{2n+3}}{Q_{2n+3}} - \frac{P_{2n+1}}{Q_{2n+1}} = \frac{1}{Q_{2n+3}Q_{2n+2}} - \frac{1}{Q_{2n+2}Q_{2n+1}} < 0$$

that the sequence (42) is a decreasing one.

Any term of the sequence (42) is greater than any term of the sequence (41). Indeed, let us examine the numbers

$$\frac{P_{2n}}{Q_{2n}} \quad \text{and} \quad \frac{P_{2m+1}}{Q_{2m+1}}$$

and let us take the odd number k to be greater than 2n and 2m+1. It follows from (33) that

$$\frac{P_k}{Q_k} > \frac{P_{k+1}}{Q_{k+1}}, \tag{43}$$

and from the fact that (41) increases and (42) decreases, it follows that

$$\frac{P_{k+1}}{Q_{k+1}} > \frac{P_{2n}}{Q_{2n}} \tag{44}$$

and

$$\frac{P_k}{Q_k} < \frac{P_{2m+1}}{Q_{2m+1}}. \tag{45}$$

Comparing (43), (44) and (45), we obtain

$$\frac{P_{2n}}{Q_{2n}} < \frac{P_{2m+1}}{Q_{2m+1}}.$$

From (33) and (34),

$$\frac{P_{n+1}}{Q_{n+1}} - \frac{P_n}{Q_n} = \frac{1}{Q_{n+1}Q_n} < \frac{1}{n^2},$$

and therefore, as n increases, the absolute value of the difference of (n+1)th and nth convergent fractions tends to zero.

From the above considerations, it is possible to conclude that the sequences (41) and (42) have the same limit, which is also, obviously, the limit of (40). This limit is called the value of the infinite continued fraction (39).

Let us prove now that any number can be the value of no more than one continued fraction. Let us take for this purpose two continued fractions w and w' (it does not matter whether they are finite or infinite).

Let q_0, q_1, q_2, ... and q'_0, q'_1, q'_2, ... be their corresponding partial denominators. We shall show that it follows from the equation w = w' that $q_0 = q'_0$, $q_1 = q'_1$, $q_2 = q'_2$, ... and so on.

Indeed, q_0 is the integral part of the number w and q'_0 is the integral part w', so that $q_0 = q'_0$. Further, the continued fractions w and w' can be represented respectively in the form

$$q_0 + \frac{1}{w_1} \quad \text{and} \quad q'_0 + \frac{1}{w'_1}$$

where w_1 and w'_1 are again continued fractions. It follows from w = w' and $q_0 = q'_0$ that $w_1 = w'_1$ also. This means that the integral parts of the numbers w_1 and w'_1 are also equal, i.e., $q_1 = q'_1$. Continuing these arguments (induction) we see that $q_2 = q'_2$, $q_3 = q'_3$, etc.

Since a rational number can always be expanded into a finite continued fraction, it follows from the foregoing proof that it cannot be expanded into an infinite continued fraction. It follows that the value of an infinite continued fraction must necessarily be an irrational number.

The theory of expansions of irrational numbers into continuous fractions represents a branch of number theory which is rich in content and interesting in its results. We shall not delve deeply into this theory, but we shall consider only one example connected with Fibonacci numbers.

<u>8</u>.   Let us find the value of the infinite
continued fraction

$$1 + \cfrac{1}{1 + \cfrac{1}{1 + \ldots}}$$

As we have proved above, this value is $\lim\limits_{n \to \infty} a_n$.
Let us calculate this limit.

As has already been established in I, section
12, u_n is the nearest whole number to $a^n/\sqrt{5}$; this
means that

$$u_n = \frac{a^n}{\sqrt{5}} + \theta_n,$$

where $|\theta_n| < 1/2$, whatever n is.

Therefore, in view of the results of section 4,

$$\lim_{n \to \infty} a_n = \lim \frac{u_{n+1}}{u_n} = \lim_{n \to \infty} \frac{\dfrac{a^{n+1}}{\sqrt{5}} + \theta_{n+1}}{\dfrac{a^n}{\sqrt{5}} + \theta_n} =$$

$$= \lim_{n \to \infty} \frac{a + \dfrac{\theta_{n+1}\sqrt{5}}{a^n}}{1 + \dfrac{\theta_n\sqrt{5}}{a^n}} = \frac{\lim\limits_{n \to \infty} \left(a + \dfrac{\theta_{n+1}\sqrt{5}}{a^n} \right)}{\lim\limits_{n \to \infty} \left(1 + \dfrac{\theta_n\sqrt{5}}{a^n} \right)}.$$

But $\theta_{n+1}\sqrt{5}$ is a bounded quantity (its absolute
value is less than 2) and a^n continues to
increase indefinitely as n tends to infinity
(because $a > 1$). This means that

$$\lim_{n \to \infty} \frac{\theta_{n+1}\sqrt{5}}{a^n} = 0.$$

Also, for the same reasons,

$$\lim_{n \to \infty} \frac{\theta_n\sqrt{5}}{a^n} = 0,$$

and we obtain

$$\lim_{n \to \infty} a_n = a.$$

The theorem that has been proved means that the ratio of neighboring Fibonacci numbers approaches a as their suffixes increase. This result can be used for the approximate calculation of the number a. (Compare the calculation of u_n in I, section 12.) This calculation produces a very small error, even when small Fibonacci numbers are taken. For example (correct to the fifth decimal place),

$$\frac{u_{10}}{u_9} = \frac{55}{34} = 1.6176,$$

and a = 1.6180. As we see, the error is less than 0.1%.

Of the errors involved in the approximate calculation of irrational numbers by convergent fractions, it turns out that the number a represents the worst case. Any other number is describable by means of its convergent fractions in some sense more exactly than a. However, we shall not stop to consider this circumstance, interesting though it be.

IV

FIBONACCI NUMBERS AND GEOMETRY

<u>1</u>. Let us divide the unit segment AB into two
parts in such a way (figure 2) that the greater
part is the mean proportional of the smaller part
and the whole segment.

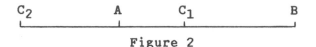

Figure 2

For this purpose, we denote the length of the
greater part of the segment by x. Obviously, the
length of the smaller part will be equal to $1 - x$
and the conditions of our problem give us the
proportion,

$$\frac{1}{x} = \frac{x}{1 - x},$$

(46)

whence

$$x^2 = 1 - x.$$

(47)

The positive root of (47) is $\frac{-1 + \sqrt{5}}{2}$ so that

the ratios in proportion (46) are equal to

$$\frac{1}{x} = \frac{2}{-1 + \sqrt{5}} = \frac{2(1 + \sqrt{5})}{(-1 + \sqrt{5})(1 + \sqrt{5})} = \frac{1 + \sqrt{5}}{2} = a$$

each. Such a division (at point C_1) is called
the <u>median</u> <u>section</u>. It is also called the <u>Golden</u>
<u>Section</u>.

If the negative value of the root of the
equation (47) is taken, the point of section C_2
lies outside the segment AB (this kind of division
is called external section in geometry), as shown
in figure 2. It is easily shown that here, too,
we are dealing with the Golden Section:

$$\frac{C_2B}{AB} = \frac{AB}{C_2A} = a.$$

2. The Golden Section appears quite frequently in geometry.

The side a_{10} of the regular decagon (figure 3) inscribed in a circle of radius R is equal to

$$2R \sin \frac{360^o}{2.10} ,$$

i.e., it is $2R \sin 18^o$.

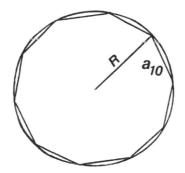

Figure 3

We now calculate $\sin 18^o$. From well known formulas of trigonometry we have

$\sin 36^o = 2 \sin 18^o \cos 18^o$,

$\cos 36^o = 1 - 2 \sin^2 18^o$,

so that

$\sin 72^o = 4 \sin 18^o \times \cos 18^o (1 - 2 \sin^2 18^o)$.   (*)

Since

$\sin 72^o = \cos 18^o \neq 0$,

then it follows from (*) that

$1 = 4 \sin 18^\circ (1 - 2 \sin^2 18^\circ)$,

and therefore $\sin 18^\circ$ is one of the roots of the equation

$1 = 4x(1 - 2x^2)$,

or

$8x^3 - 4x + 1 = 0$.

Factorizing the left-hand side of the latter equation, we obtain

$(2x - 1)(4x^2 + 2x - 1) = 0$,

whence

$$x_1 = \frac{1}{2}, \quad x_2 = \frac{-1 + \sqrt{5}}{4}, \quad x_3 = \frac{-1 - \sqrt{5}}{4}.$$

As $\sin 18^\circ$ is a positive number, other than $1/2$, therefore $\sin 18^\circ = \dfrac{\sqrt{5} - 1}{4}$.

Thus,

$$a_{10} = 2R \frac{\sqrt{5} - 1}{4} = R \frac{\sqrt{5} - 1}{2} = \frac{R}{a},$$

In other words, a_{10} equals the larger part of the radius of the circle, which has been divided by means of the Golden Section.

In practice, in calculating a_{10} we can use the ratio of neighboring Fibonacci numbers (I, section 12 or III, section 8) instead of a, and reckon approximately that a_{10} is $8/13$ R or even $5/8$ R.

<u>3</u>.   Let us examine a regular pentagon. Its diagonals form a regular pentagonal star.

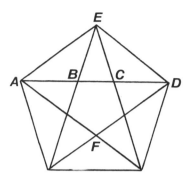

Figure 4

The angle AFD equals 108°, and the angle ADF equals 36°. Therefore, according to the sine rule,

$$\frac{AD}{AF} = \frac{\sin 108°}{\sin 36°} = \frac{\sin 72°}{\sin 36°} = 2 \cos 36° = 2\frac{1 + \sqrt{5}}{4} = a.$$

Since it is obvious that AF = AC, then

$$\frac{AD}{AF} = \frac{AD}{AC} = a,$$

and the segment AD is divided at C according to the Golden Section.

But from the definition of the Golden Section,

$$\frac{AC}{CD} = a.$$

Noting that AB = CD, we obtain

$$\frac{AC}{AB} = \frac{AB}{BC} = a.$$

Thus, of the segments BC, AB, AC, AD, each is a times greater than the preceding one.

It is left to the reader to prove that the

equality $\dfrac{AD}{AE}$ = a   also holds.

<u>4</u>. Let us take a rectangle with sides a and b and proceed to inscribe in it the largest possible squares, as shown in figure 5.

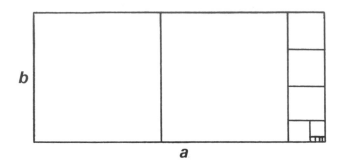

Figure 5

The arguments in II, section 2 show that such a process in the case of whole a and b corresponds to the Euclidean algorithm as applied to these numbers. The numbers of squares of equal size is in this case (III, section 1) equal to the corresponding partial denominators of the expansion of a/b into a continued fraction.

If a rectangle whose sides are to each other as neighboring Fibonacci numbers is divided into squares (figure 6), then on the basis of III, section 4, all squares except the two smallest ones are different.

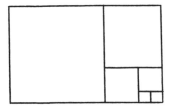

Figure 6

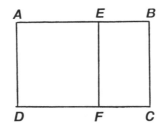

Figure 7

Now, let the ratio of the sides of a rectangle be equal to a. (We shall call such rectangles "Golden Section rectangles" for short.) We now prove that after inscribing the largest possible square into a Golden Section rectangle (figure 7), we again obtain a Golden Section rectangle.

Indeed,

$$\frac{AB}{AD} = a,$$

also, AD = AE = EF since AEFD is a square.

This means that

$$\frac{EF}{EB} = \frac{AB - EB}{EB} = a^2 - 1.$$

But $a^2 - 1 = a$, so that

$$\frac{EF}{EB} = a.$$

It is shown in figure 8 how a Golden Section rectangle can be "nearly completely" exhausted by means of squares I, II, III, ... Each successive time a square is inscribed, the remaining figure is a Golden Section rectangle.

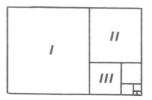

Figure 8

The reader should compare these arguments with sections 4 and 8 of the preceding chapter. We note that if a Golden Section rectangle I and squares II and III are inscribed in a square, as shown in figure 9, the remaining rectangle turns out to be a Golden Section rectangle also. The proof of this is left to the reader.

5. Golden Section rectangles seem "proportional" and are pleasant to look at. Things of this shape are convenient in use. Therefore, many "rectangular" objects of everday use (books, matchboxes, suitcases and similar things) are given this particular form.

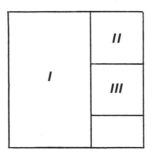

Figure 9

Various idealist philosophers of ancient and medieval times raised the outward beauty of Golden Section rectangles and other figures which conform to the rules of median section into an aesthetic and even a philosophic principle. They tried to explain natural and social phenomena in terms of the Golden Section and certain other number relationships, and they carried out all kinds of mystic "operations" on the number a and its convergent fractions. It is clear that such "theories" have nothing in common with science.

Figure 10

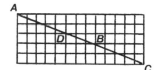

Figure 11

<u>6</u>. We shall round off our presentation with a little geometrical joke. We shall demonstrate a "proof" that 64 = 65.

To do this, we take a square of side 8 and cut it up into 4 parts as shown in figure 10. We put the parts together to form a rectangle (figure 11) of sides 13 and 5, i.e., of area equal to 65.

The explanation of this phenomenon, puzzling at first sight, is easily found. The point is that the points A, B, C and D in figure 11 do not really lie on the same straight line, but are the vertices of a parallelogram, whose area is exactly equal to the "extra" unit of area.

This plausible but misleading "proof" of a statement which is known beforehand to be incorrect (such "proofs" are called sophisms) can be carried out even more "convincingly" if we take a square of side equal to some Fibonacci number with a sufficiently large even suffix, u_{2n}, instead of a square with side 8. Let us cut up this square into parts (figure 12):

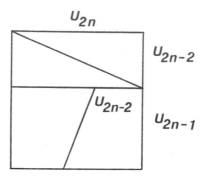

Figure 12

and let us put these parts together to form a rectangle (figure 13):

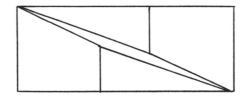

Figure 13

The "empty space" in the form of a parallelogram stretched along the diagonal of the rectangle is equal in area to unity, according to I, section 6. It is easily calculated that the greatest width of this slit, i.e., the height of the parallelogram, is equal to

$$\frac{1}{\sqrt{u_{2n}^2 + u_{2n-2}^2}} \cdot$$

If, therefore, we take a square with side 21 cm and "convert" it into a rectangle with sides 34 cm and 13 cm, the greatest width of the slit is found to be

$$\frac{1}{\sqrt{21^2 + 8^2}} \text{ cm,}$$

i.e., about 0.4 mm, which is difficult to detect by eye.

V

CONCLUSION

Not all the problems connected with Fibonacci numbers can be solved as easily as the ones we have considered. We shall indicate several problems the answers to which are either not known at all or can only be obtained by quite complicated means with the application of much more powerful methods of investigation.

<u>1</u>. Let u_n be divisible by a certain prime number p, while none of the Fibonacci numbers smaller than u_n is divisible by p. In this case we shall call the number p "the proper divisor of u_n". For example, 11 is the proper divisor of u_{10}, 17 is the proper divisor of u_9, and so on.

It turns out that any Fibonacci number except u_1, u_2, u_6 and u_{12} possesses at least one proper divisor.

<u>2</u>. The natural question arises: What is the suffix n of the Fibonacci number whose proper divisor is the given prime number p?

From II, section 13 we know that $n \leq p^2$. It is possible to prove that $n \leq p + 1$. Furthermore, it is possible to establish that if p is of the form $5t \pm 1$, then u_{p-1} is divisible by p, and if p is of the form $5t \pm 2$, then u_{p+1} is divisible by p. However, we have no formula to indicate the suffix of the term with the given proper divisor p.

<u>3</u>. We have proved in II, section 9 that all Fibonacci numbers with composite suffixes, except u_4, are composite themselves. The converse is not true, since, for example, $u_{19} = 4181 = 37 \times 113$. The question arises: is the number of all prime Fibonacci numbers finite or infinite? In other words, is there among all the prime Fibonacci numbers a greatest one? At this moment this question is still far from being solved.